AMAZING RAILROADS

Black Rabbit Books | Joanne Mattern

TABLE OF CONTENTS

1. Switzerland's Stoosbahn   4

2. Malaysia's Jungle Train   6

3. Europe's Orient Express 10

4. Colorado's Durango & Silverton 13

5. New Zealand's Raurimu Spiral 16

6. Russia's Trans-Siberian Railway 19

More to Explore   21

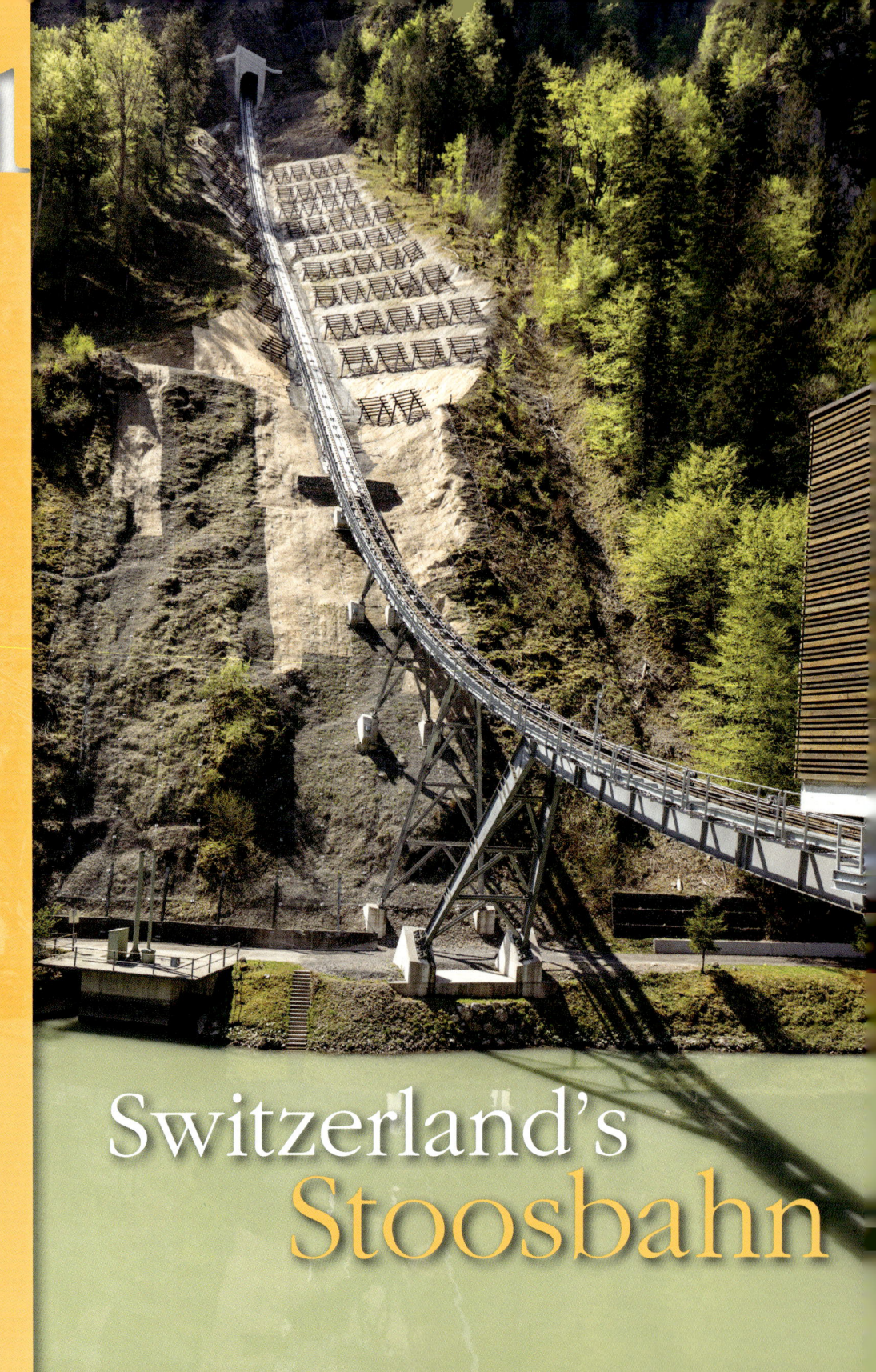

1

Switzerland's
Stoosbahn

Switzerland's Stoosbahn (STEWS-bahn) is a funicular railway. It is the steepest in the world. Two cars are connected to a cable. As one is pulled up a steep track, the other goes down. The two cars balance each other. The cable loops around a pulley at the top.

The Stoosbahn cars travel about one mile (1.6 kilometers) per hour. Each car carries passengers. The ride takes four to six minutes. It goes straight up and down a mountain. A special system keeps the glass cars from tipping.

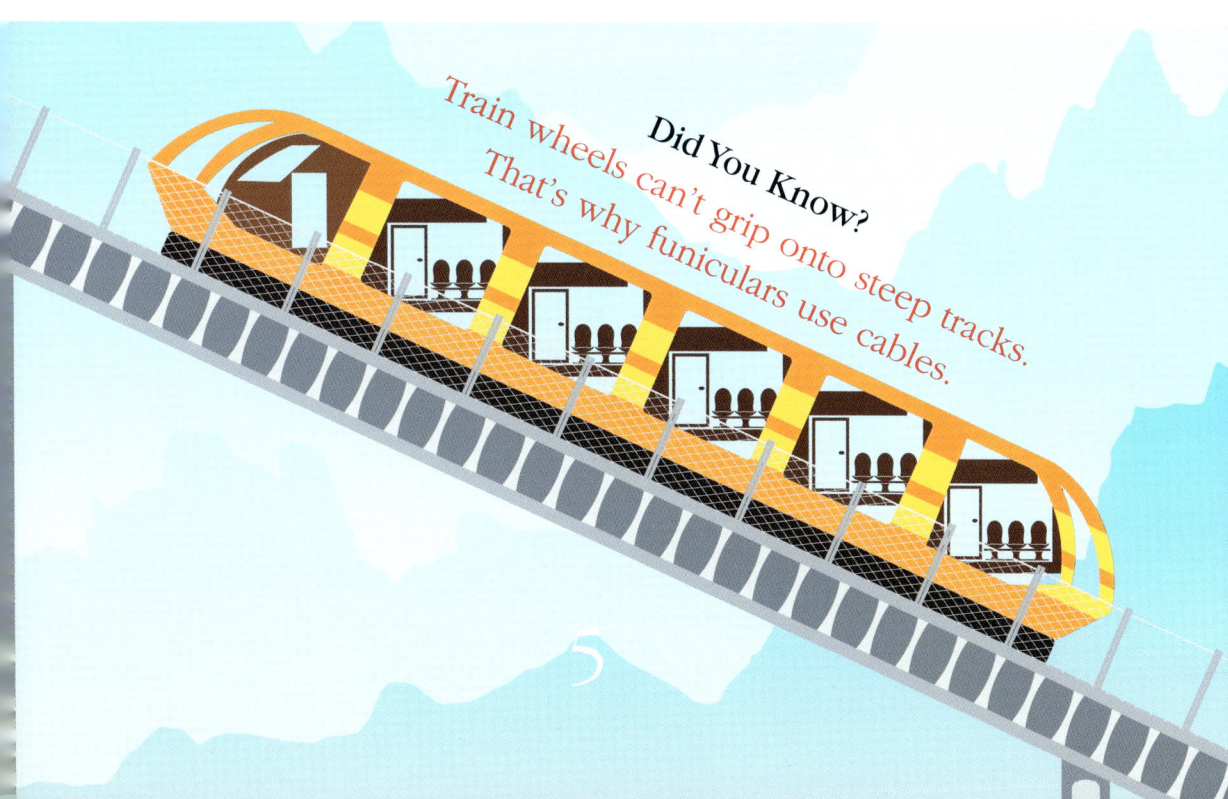

Did You Know? Train wheels can't grip onto steep tracks. That's why funiculars use cables.

Malaysia's Jungle Train

Imagine riding a train through a rainforest. People do this in Malaysia. They ride the Jungle Train. The train goes through the thick forest. The trees touch the cars on both sides.

The railway first opened in 1910. It runs through northeast Malaysia. The track is 327 miles (526 km) long. It took 21 years to build. There's just one track. Sometimes the train pulls onto a siding. Then another train can pass by.

Did You Know?
The train also passes large plantations. Palm oil is made there.

The Orient Express was once the fanciest way to travel. The train went between Istanbul and Paris. It ran from 1883 to 1977. It was more than 1,700 miles (2,740 km) long. Passengers rode in beautiful rooms. They had expensive rugs and furniture. The dining car served the best food. Kings, queens, and spies rode the Orient Express.

The Orient Express started up again in 1982. It still runs today. It travels between London and Venice.

Did You Know?
Many books and movies have been set on the Orient Express.

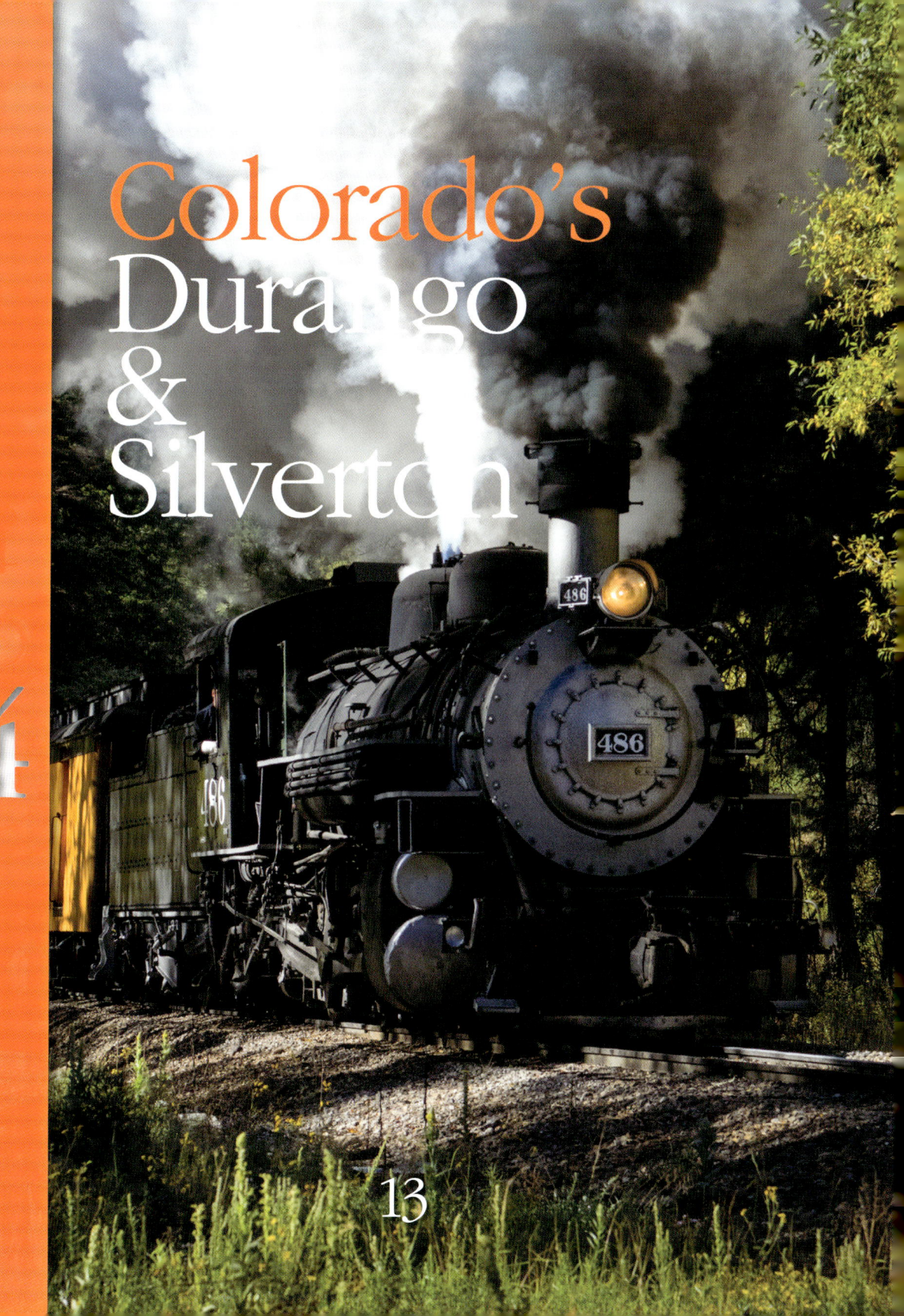

Colorado's Durango & Silverton

4

13

This train line runs between two cities. They are Durango (duh-RANG-goh) and Silverton. Both are in Colorado, United States. The train started in 1882. It was built to carry gold and silver. It also carried passengers. People loved the beautiful views.

The train has operated for over 135 years. People still enjoy riding. It travels through the San Juan National Forest. It also goes through mountains. Steam **locomotives** pull the cars. They travel on narrow tracks. The train's top speed is 18 miles (29 km) per hour.

Think About It
Why do you think people enjoy traveling an old-fashioned way?

New Zealand's Raurimu Spiral

New Zealand is an island east of Australia. The country needed a railway between two towns. But the land between them was very steep. There was an **engineer** named Robert Holmes. He came up with a plan. His railway would include curves, tunnels, and a circle. Trains wouldn't have to go straight up or down a hill.

The Raurimu Spiral (rowr-EE-moo SPY-ruhl) was built. It opened in 1908. It became an engineering historic site in 2012.

Model of the Raurimu Spiral train track in New Zealand.

Think About It
What problems do engineers face when planning railroads?

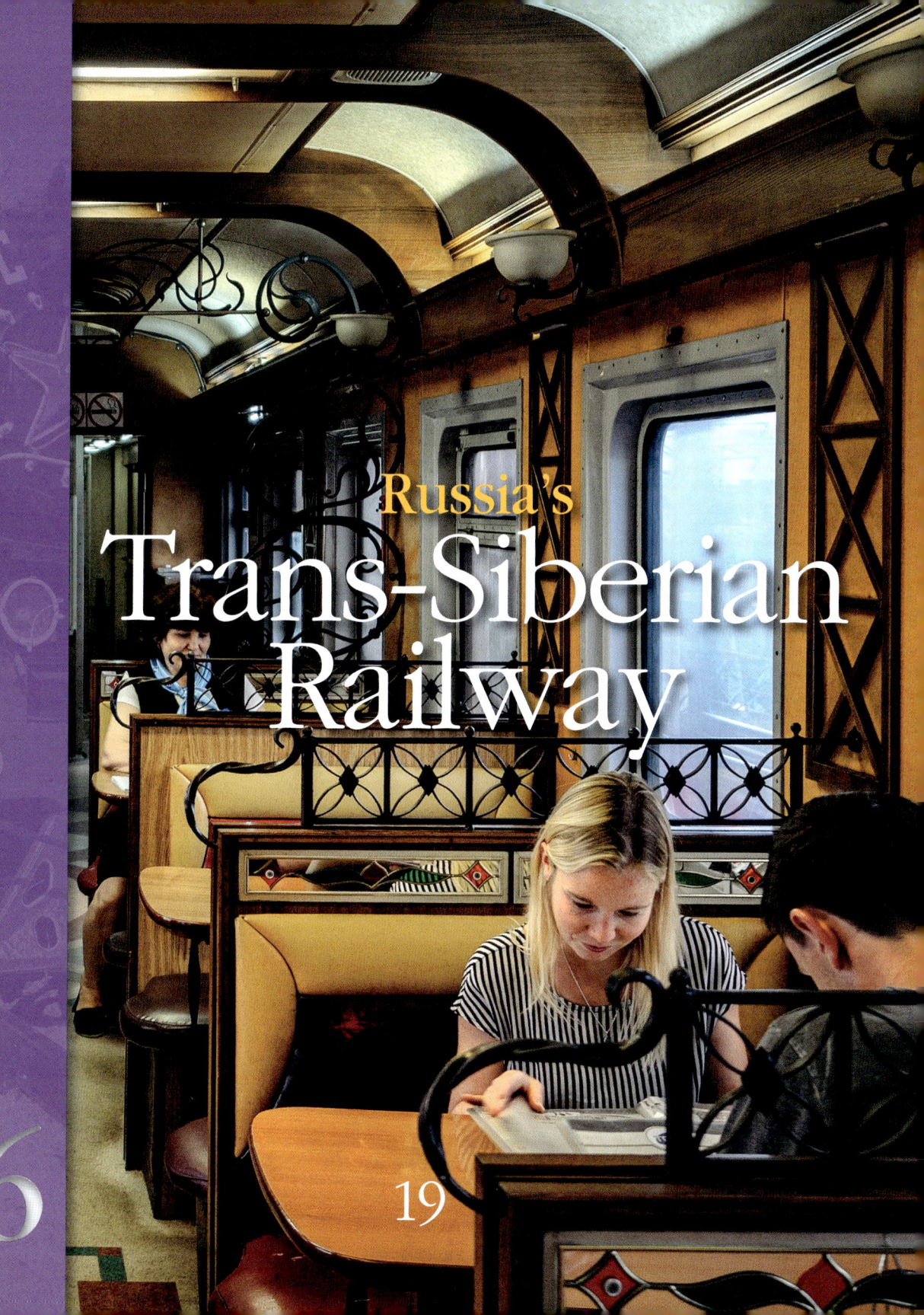

Russia's Trans-Siberian Railway

Passengers love the Trans-Siberian (TRANZ-sahy-BEER-ee-uhn) Railway. It takes them on an epic journey. This railway is the longest in the world. It stretches 5,778 miles (9,299 km). It goes from Moscow to Vladivostok. This is in Russia's Far East.

The railway was finished in 1916. Today, the trip takes about six days. Passengers travel through eight time zones. There are many stops. People see big cities. They pass rural villages. Views include natural wonders too. The Ural Mountains and Lake Baikal are nearby.

This railway is longer than the Great Wall of China.

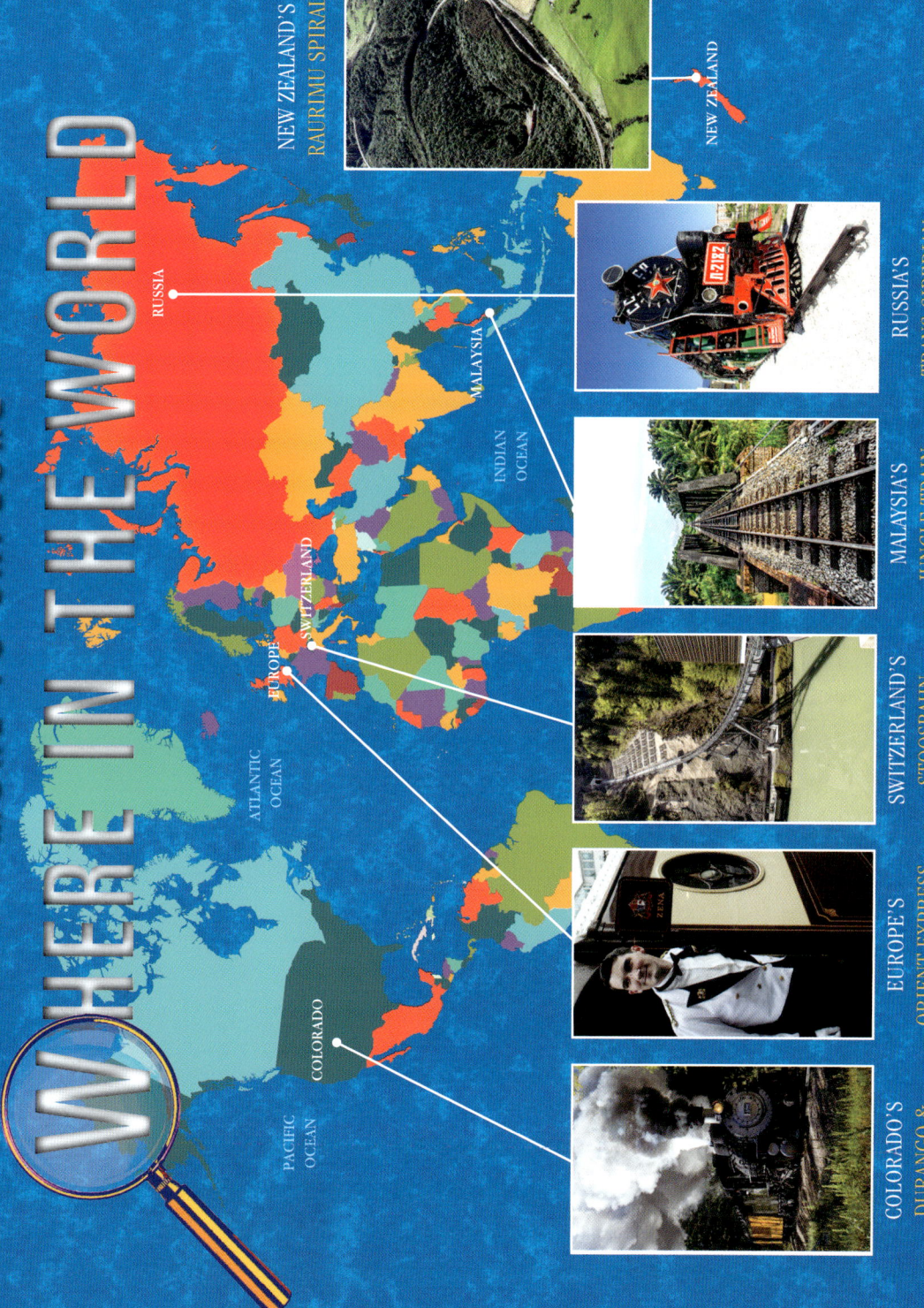

MORE TO EXPLORE
FANTASTIC FACTS

It took 14 years to plan and build the Stoosbahn.

The Jungle Train sometimes stops and waits for another train to pass.

It costs about $4,500 to ride the Orient Express.

The Orient Express didn't run during World War I and World War II.

The Trans-Siberian Railway passes through 87 cities and towns.

It was dangerous for the workers on the **Trans-Siberian Railway**. They had to worry about tigers, floods, and terribly cold weather.

MORE TO EXPLORE
COOL COMPARISONS

How do these amazing railroads compare?

Stoosbahn: 1.1 miles (1.8 km) long

Raurimu Spiral: 4.5 miles (7.2 km) long

Durango & Silverton: 45.4 miles (73 km) long

Jungle Train: 327 miles (526 km) long

Orient Express: 1,700 miles (2,740 km) long

Trans-Siberian Railway: 5,778 miles (9,299 km) long

MORE TO EXPLORE
RESOURCES

Glossary

engineer (en-juh-NEER) A person who designs and builds complicated things.

funicular (fyoo-NIK-yuh-ler) A railway going up and down a mountain that carries people in cars pulled by a moving cable.

locomotive (loh-kuh-MOH-tiv) The vehicle that produces the power that pulls a train.

plantation (plan-TEY-shuhn) A large area where crops are grown.

rural (ROOR-uhl) Having to do with the countryside.

siding (SAHY-ing) A short railway track connected to a longer track.

steep (STEEP) Rising or falling sharply.

Read More

Duling, Kaitlyn. *Trains*. Minneapolis: Bellwether Media, Inc., 2023.

Klepeis, Alicia. *Superfast Trains*. Minneapolis: Jump!, 2022.

Index

cables, 5
engineers, 17, 18
forests, 6, 14
funicular, 5
historic sites, 17
mountains, 5, 14, 20
passengers, 5, 11, 14, 20
railroad lengths, 6, 11, 20, 23
speed, 5, 14

TOP RANK is published by Black Rabbit Books, P.O. Box 227, Mankato, MN 56002 • COPYRIGHT © 2025 Black Rabbit Books. All rights reserved. No part of this book may be reproduced in any form without written permission from the publisher. • Top Rank is an imprint of Black Rabbit Books. • Edited by Alissa Thielges • Designed by Danny Nanos • Photographs © Alamy: Phil Crean A, 17; Dreamstime: Rodrigolab, 19; Flickr: Jenny Scott, 16; Getty: olu, 11, Fotos by Fudge, 13, 21, Manchester Daily Express, 12, Martin Godwin, 10, Murmakova, 4, 21; Public Domain: 14; Shutterstock: ABCDstock, 20, Bojan Pesic, 18, Fotos by Fudge, cover, 21, Layne V. Naylor, 15, Magdalena Cvetkovic, 23, May_Lana, 6–7, 8–9, 21, seeyah panwan, 21, Umlaut1968, 2–3, vectorOK, 5, William Cushman, cover • Printed in the United States of America
Library of Congress Cataloging-in-Publication Data: Names: Mattern, Joanne, 1963- author. | Title: Amazing railroads / by Joanne Mattern. | Description: Mankato, MN: Black Rabbit Books, [2025] | Series: Design marvels | Audience: Ages 8–11 | Audience: Grades 4–6 | Identifiers: LCCN 2023058230 | ISBN 9781632357915 (library binding) | ISBN 9781645820703 (ebook) | Subjects: LCSH: Railroads—Juvenile literature. | Railroads—Design and construction—Juvenile literature. | Railroad trains—Juvenile literature. | CYAC: Railroads. | Railroad trains. | LCGFT: Instructional and educational works. | Classification: LCC TF148 .M3795 2025 | DDC 625.1—dc23/eng/20240129 | LC record available at https://lccn.loc.gov/2023058230